Recommendations for Developing Space Suit Integrated Food Systems and Delivering Before, During, and After Lunar EVA

Lichar Dillon, PhD

NIMBLE BOOKS LLC: THE AI LAB FOR BOOK-LOVERS

~ FRED ZIMMERMAN, EDITOR ~

Humans and AI making books richer, more diverse, and more surprising.

PUBLISHING INFORMATION

(c) 2024 Nimble Books LLC
ISBN: 978-1-60888-287-8

AI-GENERATED KEYWORD PHRASES

space suit integrated food systems; lunar extravehicular activities; in-suit nutrition; space programs; nutrition support; Artemis program; EVAs; EVA nutrition support; commercial off-the-shelf (COTS) foods; in-suit nutrition delivery concept designs; suit placement considerations; astronaut feedback and recommendations; 400-600 kcal of in-suit nutrition during EVAs; food systems for space flight requirements.

PUBLISHER'S NOTES

If you are old enough to remember Tang™ or Space Food Sticks from Pillsbury, you will completely understand my motivation for publishing this document. Astronauts in spacesuits are still astronauts that need to eat. The more astronauts wear spacesuits, the more they need food that can get through the spacesuit somehow. What was old, will be new once again.

This 2023 presentation offers NASA's recommendations for developing space suit integrated food systems and delivering nutrition during lunar extravehicular activities (EVAs). It addresses the critical need for adequate nutrition support during the Artemis program, which aims to conduct a higher tempo and frequency of EVAs than any previous space program. By reading this document, readers can gain insights into the challenges and potential solutions for providing in-suit nutrition to astronauts, which is crucial for their health, performance, and mission success in future space exploration endeavors.

This annotated edition illustrates the capabilities of the AI Lab for Book-Lovers to add context and ease-of-use to manuscripts. It includes several types of abstracts, building from simplest to more complex: TLDR (one word), ELI5, TLDR (vanilla), Scientific Style, and Action Items; essays to increase viewpoint diversity, such as Grounds for Dissent, Red Team Critique, and MAGA Perspective; and Notable Passages and Nutshell Summaries for each page.

ANNOTATIONS

ABSTRACTS

TL;DR (ONE WORD)

Recommendations.

EXPLAIN IT TO ME LIKE I'M FIVE YEARS OLD

This document is about how astronauts can eat while they are wearing their space suits during spacewalks. It talks about why it is important for astronauts to have enough food during these spacewalks and how they can get that food. It also discusses different ideas for how to give astronauts their food while they are in their suits. The document also includes suggestions from astronauts on what can be improved in the future. Overall, it says that astronauts need to have enough food during spacewalks and that the food

TL;DR (VANILLA)

This document provides recommendations for developing food systems for astronauts to consume during lunar extravehicular activities (EVAs). It discusses the background of in-suit nutrition, the motivation to include nutrition in space suits, and the importance of adequate nutrition support during the Artemis program. The document addresses questions about EVA nutrition support, the amount of nutrition needed during EVAs, and limitations and capabilities of space suits. It explores using commercial off-the-shelf foods, concept designs for in-suit nutrition delivery, and suit placement considerations. Astronaut feedback and recommendations are also included. Overall, it emphasizes the need for 400-600 kcal of in-suit nutrition during EVAs and developing food systems that meet safety, stability, and acceptability requirements.

SCIENTIFIC STYLE

This document provides recommendations for developing space suit integrated food systems and delivering nutrition before, during, and after lunar extravehicular activities (EVAs). The document discusses the background of in-suit nutrition in previous space programs and the motivation to reintroduce nutrition support during EVAs. With the Artemis program targeting a higher tempo and frequency of EVAs, it is crucial to provide adequate nutrition support. The document addresses general questions related to EVA nutrition support, such as the amount of nutrition needed during EVAs and the limitations and functional capabilities of space suits. It explores the use of commercial off-the-shelf (COTS) foods, concept designs for in-suit nutrition delivery, and suit placement considerations. Astronaut feedback and recommendations for future work are also provided. Overall, the document emphasizes the need for at least 400-600 kcal of in-suit nutrition during EVAs and highlights the importance of developing food systems that meet space flight requirements in terms of safety, stability, and acceptability.

ACTION ITEMS

Conduct further research on the nutritional needs of astronauts during EVAs and the potential impact of inadequate nutrition on their performance and health.

Explore different options for delivering nutrition during EVAs, such as squeeze pouches, bite-sized snacks, or liquid meals.

Collaborate with food scientists and engineers to develop space suit-integrated food systems that meet the requirements for safety, stability, and acceptability.

Conduct usability testing with astronauts to gather feedback on the functionality and comfort of different in-suit nutrition delivery systems.

Consider the logistics of storing and replenishing in-suit nutrition supplies during long-duration

VIEWPOINTS

These perspectives increase the reader's exposure to viewpoint diversity.

GROUNDS FOR DISSENT

A member of the organization responsible for this document might have principled, substantive reasons to dissent from this report. Here are a few possible reasons:

Resource Allocation: The dissenting member may argue that allocating significant resources towards developing space suit integrated food systems is not a priority at this time. They might believe that resources should be directed towards other critical areas of space exploration, such as propulsion systems or life support technologies. They could advocate for redirecting funds and efforts to these more pressing needs instead.

Cost-Effectiveness: Another reason for dissent could be the perceived lack of cost-effectiveness in developing and implementing in-suit nutrition systems. The dissenting member might argue that the expenses involved in researching, designing, and testing these systems outweigh the potential benefits. They may suggest alternative methods of providing nutrition during EVAs that are more cost-effective and efficient.

Operational Complexity: A dissenting view could stem from concerns about the operational complexity of integrating food systems into space suits. The member might argue that introducing additional components or mechanisms into already complex space suits could increase the risk of failure or compromise other functionalities. They may propose simpler alternatives or question whether the added complexity is necessary.

Safety Risks: Dissenters might raise concerns about potential safety risks associated with in-suit nutrition delivery. They could argue that any malfunction or contamination of the food system during EVAs could pose a serious threat to astronaut health and mission success. The dissenting member might propose focusing on alternative methods of meeting nutritional needs that do not involve potential safety hazards.

Astronaut Autonomy: Some members might oppose the idea of in-suit nutrition delivery due to concerns related to astronaut autonomy and decision-making. They may argue that allowing astronauts to choose and consume their meals separately from their space suits would provide them with more control over their own nutrition intake. This dissenting view could focus on preserving individual choice and preferences rather than implementing a standardized in-suit nutrition system.

Ultimately, these dissenting views challenge the prioritization, cost-effectiveness, operational complexity, safety risks, and astronaut autonomy associated with the recommendations outlined in the document. They propose alternative approaches or question the necessity of implementing in-suit nutrition systems during lunar EVAs.

Red Team Critique

Overall, the document provides a comprehensive overview of the need for in-suit nutrition during lunar extravehicular activities (EVAs) and addresses various considerations related to developing space suit integrated food systems. However, there are several areas where the document could benefit from further analysis and improvement.

Firstly, while the document discusses the motivation to put nutrition back in space suits, it does not sufficiently justify this decision. It would be helpful to provide more context on why previous space programs did not prioritize in-suit nutrition and what new developments have prompted reconsideration. Additionally, it would be beneficial to include a discussion on any potential risks or challenges associated with integrating food systems into space suits.

Furthermore, while there is a mention of general questions related to EVA nutrition support, these questions are not explicitly stated or explored in detail within the document. It is important to identify these specific questions and address them thoroughly as they will play a crucial role in informing decisions regarding nutrition support during EVAs.

The exploration of commercial off-the-shelf (COTS) foods and concept designs for in-suit nutrition delivery is commendable. However, the document lacks an evaluation of their feasibility and effectiveness. It would be valuable to conduct a comparative analysis of different COTS foods and concept designs based on factors such as nutritional content, shelf-life, packaging efficiency, weight considerations, ease of eating in zero gravity environments etc., before making recommendations.

Additionally, more attention should be given to suit placement considerations for food storage within space suits. The document briefly mentions this aspect but fails to provide concrete recommendations or suggestions for optimal suit placement strategies that maximize accessibility without compromising mobility or safety concerns.

Although astronaut feedback and recommendations are provided at some point within the document structure-wise; however we fail see usage/usefulness/relevance beyond mentioning them if additional details about their experience/feedbacks were shared along with how these inputs influenced/formulated future plans giving true value-addition insights.

Finally, while the document emphasizes the need for at least 400-600 kcal of in-suit nutrition during EVAs, it does not provide a clear justification or scientific basis for this range. It would be beneficial to include information on how this calorie requirement was determined and whether it varies depending on factors such as duration of EVA or intensity of physical activity.

In conclusion, while the document provides valuable insights into the development of space suit integrated food systems for lunar EVAs, there are areas where further analysis and improvement can enhance its overall effectiveness. Additional justifications, thorough exploration of questions related to EVA nutrition support, evaluation of COTS foods and concept designs, concrete recommendations for suit placement considerations and more detailed analysis of astronaut feedback would significantly strengthen the document's content. Furthermore; providing a clearer scientific basis for calorie requirements would also improve understanding and decision-making in this area.

MAGA PERSPECTIVE

This document is just another example of the liberal agenda infiltrating our space programs. Instead of focusing on important issues like American exceptionalism and putting America first, they are wasting time and resources on developing food systems for astronauts. This is just another way for them to push their socialistic ideas of redistributing resources and providing for everyone's needs.

The notion that nutrition needs to be provided during EVAs is absurd. If these astronauts can't go a few hours without eating, then maybe they're not cut out for space exploration in the first place. We should be selecting strong, self-reliant individuals who can handle the physical demands of the job, not coddling them with in-suit nutrition.

Furthermore, the idea of using commercial off-the-shelf (COTS) foods is just another attempt to cater to mainstream consumerism rather than focusing on American innovation. Our astronauts should be fed with top-of-the-line, specially designed meals that represent the best of what our country has to offer.

It's also concerning that the document mentions astronaut feedback and recommendations for future work. Who cares what these astronauts think? They are just government employees doing a job, not experts in nutrition or food systems. Their opinions should not be given any weight or consideration when making decisions about how we allocate resources.

In conclusion, this document is a prime example of the left's misguided priorities and disregard for true American values. Instead of wasting time on in-suit nutrition, we should be focusing on advancing our space program in a way that benefits our nation and puts America first.

Page-by-Page Summaries

BODY-1   Recommendations for developing space suit integrated food systems and delivering nutrition before, during, and after lunar EVA.

BODY-2   In-suit nutrition was implemented in the Apollo program but discontinued in the Space Shuttle due to various reasons. The motivation to reintroduce it is being discussed at the 2023 Human Research Program Investigators Workshop.

BODY-3   The Artemis program aims to have a high frequency of extravehicular activities (EVAs). Adequate nutrition support, including in-suit options, is crucial. Meeting caloric needs before, during, and after EVAs should be considered. However, the ability to provide additional nutrition for EVA performance is limited.

BODY-4   The page discusses questions regarding nutrition support, food formulations, suit limitations, crewmember capabilities, and astronaut practices for extravehicular activities (EVA) in space.

BODY-5   The page discusses the energy intake needed for enhanced performance during spacewalks, suggesting a target of consuming at least 400 kcal in the suit to support performance. Consuming more than 600 kcal is not expected to provide additional benefits.

BODY-6   COTS foods tested for space exploration did not meet microbiological and foreign object debris requirements, highlighting the need for further assessments. Nutritional content, spatial evaluation, and sensory evaluation are important factors to consider in selecting safe and acceptable products for space missions.

BODY-7   Concept designs for pre-filled bags, food sticks, hydratable bags, and helmet ports for liquid or gel substances. Contact information provided for Lichar Dillon at NASA.

BODY-8   Suit placement considerations for the Exploration Extravehicular Mobility Unit (xEMU) are discussed.

BODY-9   This page provides an example of volumetric analysis for positioning a drink bag on the xEMU hard upper torso, highlighting limitations in available volume.

BODY-10   Astronaut feedback on hydration, nutrition, and waste management in space suits. Ready-to-consume drink bags were preferred, but limited shelf-life was a concern. Some astronauts desired solid food options. Feasibility of helmet feed port uncertain. Keeping food outside of the suit was seen as positive. Multiple systems should be considered.

BODY-11   Future pressure suits should be designed to accommodate at least 400 kcal of food for individuals inside the suit. The vehicle food system should provide the remaining energy before or after suited operations. Considerations include location, volume, hands-free access, and overcoming pressure differentials during consumption.

BODY-12   Future work includes expanding testing and development of space flight foods, incorporating prototype EVA food systems in realistic scenarios, and involving crew members in the development process.

BODY-15   The page provides contact information for Lichar Dillon and mentions the 2023 Human Research Program Investigators Workshop.

BODY-16   Future Artemis missions require space suits with in-suit nutrition systems to support astronauts during longer EVAs. This project recommends providing 400-600 kcal of nutrition within the suit, either

NOTABLE PASSAGES

BODY-3   *"The Artemis program is targeting a higher tempo and frequency of extravehicular activities (EVAs) than any previous space program."*

BODY-5   *"Recommendation: at least 400 kcal in the suit to support performance"*

BODY-6   *"Space exploration is not the place to get food poisoning."*

BODY-10 *"Concept of keeping food outside of the suit was regarded as a positive attribute."*

BODY-16 *"The ability to meet the increased need for nutrition during surface EVAs through provision of nutrients in the suited configuration would benefit overall crew health, performance, and morale, and thus increase the likelihood of mission success."*

BODY-17 *"Major Gap – NO current COTS products and delivery systems exist that meet requirements"*

Background (Apollo/Shuttle)

- The concept of in-suit nutrition is nearly as old as the space program itself

- In-suit Nutrition Implemented (Apollo)
  - Food Stick
    - Individual preferences (some liked it, some did not)
  - Flavored (K-fortified) Drink
    - Aimed to improve hydration, little nutritional value other than Potassium
  - Helmet Port
    - For contingency use only (never used in-mission), difficult to use

- In-suit Nutrition Discontinued (Space Shuttle)
  - Fruit 'roll-up' stick
    - Reasons for discontinuation: messy, crew consumed before extravehicular activity (EVA), not mission critical

- What is the motivation to put nutrition back in space suits?

Artemis

- The Artemis program is targeting a higher tempo and frequency of extravehicular activities (EVAs) than any previous space program
  - 4 EVAs within 5 days

- Providing adequate nutrition support is critical
  - In-suit nutrition options should be included as part of the overall food systems in a strategy to meet crewmember needs
  - Consideration to meeting EVA caloric needs should include options to fuel before, during, and after EVA

- The ability to provide additional nutrition to support EVA performance is limited
  - In-suit nutrition should benefit the astronaut during a single EVA

General Questions

- How much and what type of EVA nutrition support should be provided in the suit and how much can be leveraged by supplementing rations before and after the EVA?

- What food formulations meet partial gravity constraints and are appropriate and safe for in-suit operations?

- What are the limitations of the suit?

- What are the functional capabilities of a suited crewmember?

- What are practices and preferences that astronauts commonly follow that should be considered?

How much is enough?

- Food system must provide addition of 200 kcal/hr EVA (x 8 hours = 1600 kcal/day)
  - Not all of this needs to be provided in the pressure suit
  - Some can be provided pre/post EVA
  - How much is necessary for enhanced performance?

- Draw from sports physiology and military literature
  - Example: Energy intake to support expenditure during a 10-hour exercise trial*
  - Approximately 21 g carbohydrate (CHO) intake per hour of exercise (~84 kcal/hr)

- Expenditure is expected to be lower during EVA
  - Reasonable target is to consume ~50-80 kcal per hour of EVA (x 8 hours ≥ 400 kcal in the suit)

- Recommendation: at least 400 kcal in the suit to support performance
  - As CHO or the equivalent energy balanced as CHO/fat/protein
  - Improve palatability, acceptability, muscle recovery, etc.
  - Providing more than ~600 kcal in the suit is not expected to add benefit to performance
  - Waste if not consumed

* Harger-Domitrovich SG, McClaughry AE, Gaskill SE, Ruby BC. Exogenous carbohydrate spares muscle glycogen in men and women during 10 h of exercise. Med Sci Sports Exerc. 2007;39(12):2171-9.

BODY-5

Commercial off the Shelf (COTS) Foods

- COTS bars and beverage powders were identified that met minimum nutritional requirements
  - Only if multiple servings are used

- Subset of COTS products tested
  - Did not meet both microbiological (rehydrated powders) and foreign object debris (FOD; solid food bars) requirement.
  - → Limited (2-hour) window of consumption for liquid nutrition following breaking of the packaging seal
  - In agreement with FDA guidelines
  - Space exploration is not the place to get food poisoning

- Additional assessments required for any product
  - Nutritional content
    - Ensure nutrients remain within requirements of the whole diet.
  - Spatial evaluation
    - Configuration, FOD compliance, and delivery potential within the suit.
  - Sensory evaluation
    - Verify that selected products remain safe and acceptable for the duration of the EVA mission.

Examples

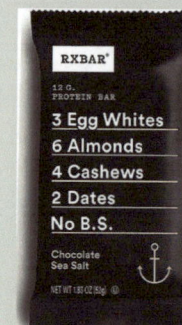

Lichar Dillon | edgar.l.dillon@nasa.gov

2023 Human Research Program
Investigators Workshop

Concept Designs

Pre-filled Bag
(Liquid)

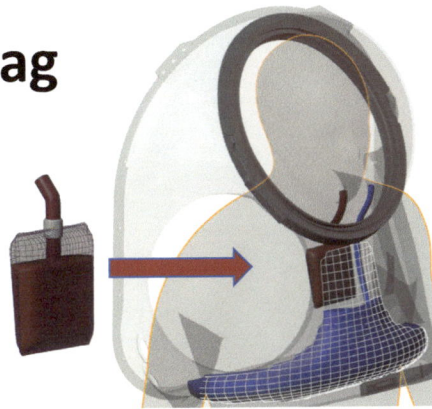

Hydratable Bag
(Liquid + Powder)

Food Stick
(Solid)

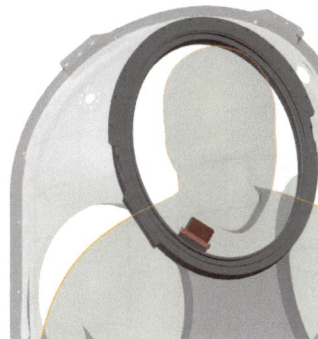

Helmet Port
(Liquid or Gel)

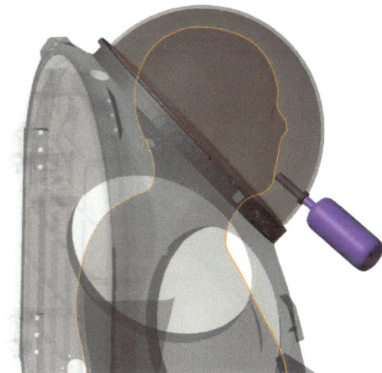

Suit Placement Considerations

- Exploration Extravehicular Mobility Unit (xEMU) Example

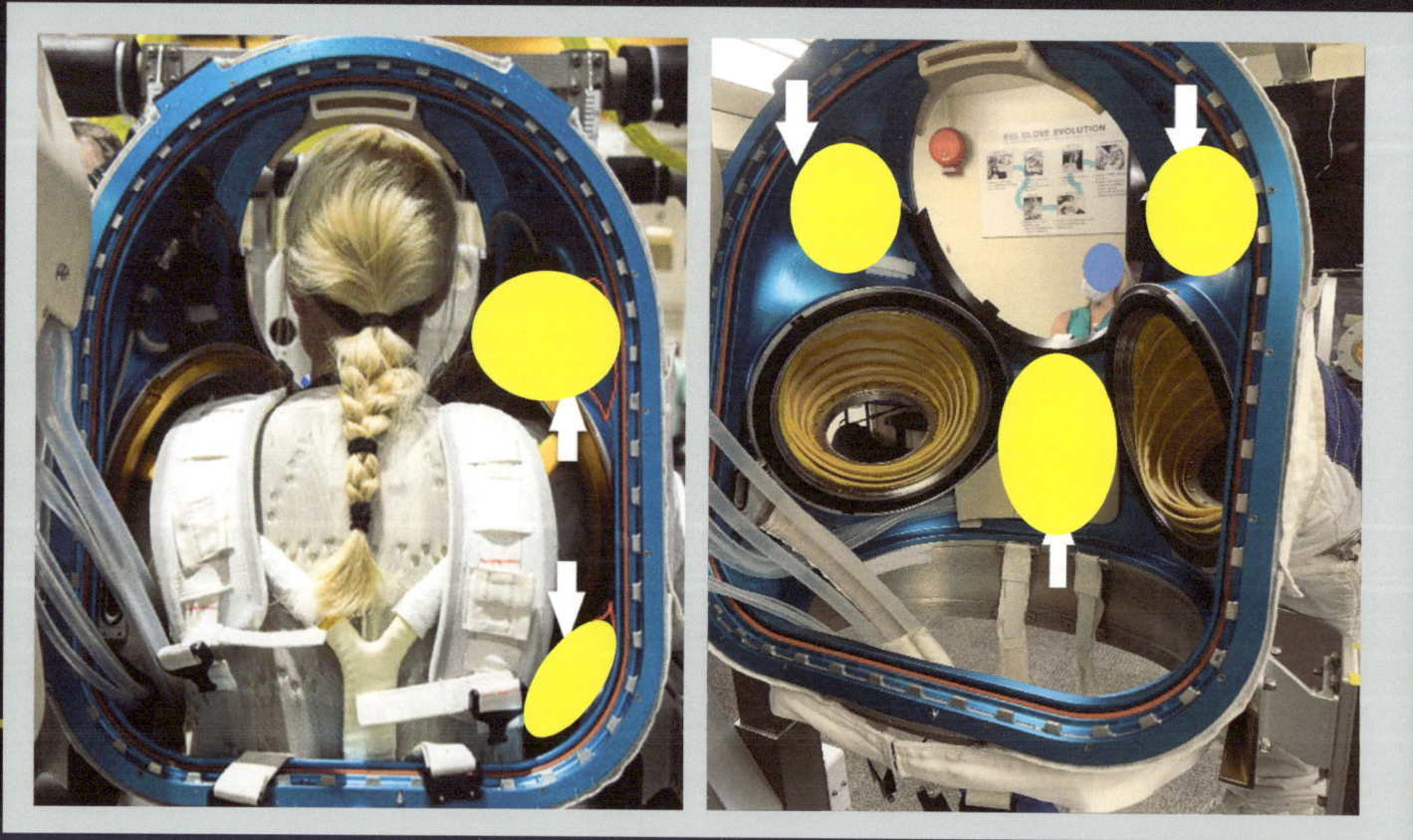

Volumetric Analysis Example

- Preliminary positioning of a drink bag on top of the DIDB
  - Mid-chest of current xEMU hard upper torso (HUT) designs

- Using population averages as female and male models
  - xEMU HUT and the disposable in-suit drink bag (DIDB) filled to capacity (1070 mL)
  - Maximum volume available for a median size female: 260 ml
  - Maximum volume available for a median size male: 190 ml

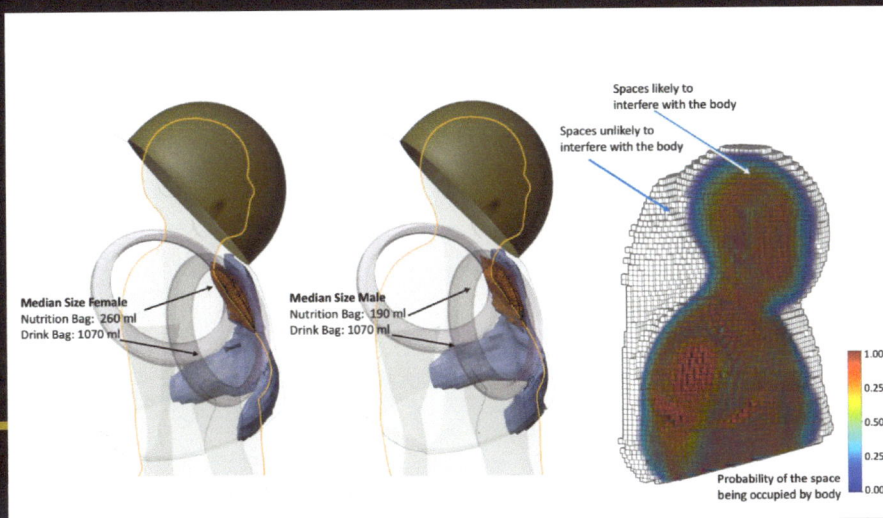

Working assumptions (1 kcal·mL^{-1}) demonstrates limitations at this location (400 mL required for liquids)

Astronaut Feedback

- 2 Qualtrics based questionnaires for Crew Feedback (rolled out separately)
  - General Preferences for Hydration, Nutrition, and Waste Management in Space Suits (n=25)
  - Concept Design References for In-Suit Nutrition Systems (n=17, 9 with EVA experience)

Summary (Feedback consistent with input from project SMEs)

- Ready to consume drink bags received the highest ranking (but not unanimous)
  - Limited shelf-life reduced enthusiasm for the hydratable drink bag option.

- Solid food as a high-density food option was desired by some.
  - Lack of experience & concern of messiness may have contributed to lower enthusiasm.

- Uncertain feasibility of implementing a helmet feed port
  - Concept of keeping food outside of the suit was regarded as a positive attribute

- Multiple systems should be considered

Summary Recommendations

- Future pressure suit designs should accommodate provision of at least 400 kcal to the individual confined to the suit.
  - The vehicle food system needs to provide the remaining energy before or after suited operations during regular meals.
  - COTS or developed foods
  - CHO/Fat/Protein

- Initial xEMU location considered (front of chest near the DIDB) is currently volumetrically limited to approximately 200–250 mL
  - This volume is <u>insufficient</u> to provide 400 kcal as liquid (~400 mL)
  - Other locations are possible, but were not analyzed
  - Future suit development to consider minimum of 400 mL for in-suit liquids

- Hands-free access
  - Should not interfere with jaw, head, neck motions or field of view
  - Capability to break the seal of prefilled food grade quality packaging (liquids/gels)

- Helmet feed ports
  - Requires staging of the foods prior to the EVA
  - Overcoming pressure differentials during consumption of external foods
  - Requires physiochemical (dust) mitigation

Future Work

- Expanded COTS testing and/or development of foods that meet all space flight requirements (nutritional content/ water activity/ microbiology/ FOD/ acceptability/ etc.)
  - Continue development of multiple solutions
    - Both solid and liquid
  - Continued/parallel effort at NASA in addition to xEVAS contractor

- Incorporate prototype EVA food system within environments that closely simulate realistic EVA scenarios to evaluate human factors, ergonomics, and human-system integration of system prototypes
  - Using 3D printed mockup suit components - Physical and Cognitive Exploration Simulations (PACES)
  - EVA training runs in the NBL, and/or Active Response Gravity Offload System (ARGOS)

- End-user (Crew) involvement during development

BODY-12

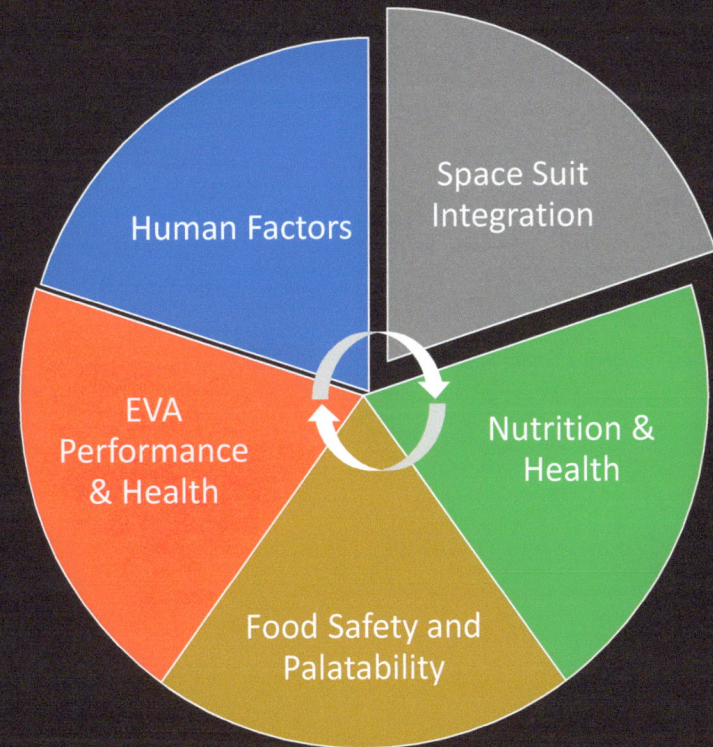

Human
Health and
Performance

NASA JOHNSON SPACE CENTER

Exploring Space | Enhancing Life

Thank you

Image: Gulf Coast at Night (NASA, International Space Station, 08/09/14)

BACKUP

ABSTRACT

NASA

INTRODUCTION

- Artemis missions will include a higher tempo and frequency of extravehicular activities (EVAs) than any previous space program. Because of the physical demands expected from the crew, future space suit designs are required to incorporate nutritional support to the astronauts during lunar surface EVAs lasting longer than 4 hours. The purpose of this project was to provide recommendations to aid the development of an in-suit system that can adequately, safely, and acceptably deliver nutrition to a crewmember while confined to a space suit during EVA.

METHODS

- Physiological, logistical, and engineering aspects of potential in-suit nutrition approaches were assessed through literature reviews, assessments of commercial off the shelf (COTS) foods, suit volumetric modeling, and feedback from subject matter experts and crewmembers. Key driving factors in the development of in-suit nutrition requirements included how much and what type of nutrition should be included, what food formulations are appropriate and safe, what are inherent limitations of space suits, what are the potential risks to the crewmember in the suit, and what practices and preferences from astronauts should be considered. Design references were conceptualized and assessed for strengths and limitations as potential in-suit nutrition systems for surface EVA.

RESULTS

- Acute exogenous energy demands vary greatly depending on activity intensity and duration, and partial energy replenishment (i.e., 60–80 kcal·hr-1 of EVA, or 460–680 kcal for EVAs lasting up to 8 hours) during activities could improve performance, safety, and recovery. COTS foods capable of providing these energy requirements exist; however, no COTS foods have been identified that pass NASA flight standards for microbiological safety and stability. In-suit nutrition delivery design references that were considered included in-suit concepts for a prefilled drink bag, a hydratable drink bag, and a solid food stick. In addition, a helmet feed port concept was considered for use with drink bags external to the suit. Volumetric models of the in-suit drink bag concepts, based on xEMU dimensions, indicate challenges of fitting formulations > 200 ml (equating to approximately 200 kcal). Astronaut feedback on the four concepts indicated that despite some individual preferences for inclusion of solid foods and helmet port designs, the prefilled drink bag concept was the most preferred. A prefilled drink bag can only be used if food safety and stability can be ensured, possibly requiring advancements in food delivery hardware.

CONCLUSION

- The ability to meet the increased need for nutrition during surface EVAs through provision of nutrients in the suited configuration would benefit overall crew health, performance, and morale, and thus increase the likelihood of mission success. It is recommended that in-suit nutrition capabilities provide at least 400–600 kcal within the suit during EVAs lasting > 4 hours and that suit designs include a dedicated volume for food grade nutrition systems. The developed food system should either allow for 1) installation of prefilled (sealed sterile) liquid nutrition in the suit and provide a mechanism to break the seal at the time that consumption is desired or 2) demonstrate that the unsealed food product shelf life allows for safe consumption after at least 12 hours of EVA.

Full report (NASA/TP-20220019344) available at https://ntrs.nasa.gov/

In-Suit Nutrition Recommendations

- Minimum of 400 kcal needed for In-Suit Nutrition
  - Minimum amount determined necessary for performance improvement
  - Food System Requirement to Supply Additional 1600 kcal for 8-hr EVA
    - Not all kcal need to be delivered in the suit, but sets a clear upper limit
    - Range of 400-600 kcal should be considered
  - Gap - xEMU primary location limited to ~200mL.
    - Future suit development to consider minimum of 400 mL for in-suit liquids.
  - Major Gap – NO current COTS products and delivery systems exist that meet requirements
    - Example: mixing powdered meal replacement with water prior to the EVA would require consumption within 2 hours of breaking package seal, thus necessitating consumption at the very beginning on an EVA
    - Risk (Foodborne Illness): 5x5 to 3x4 LxC

- Four Different Reference Delivery Systems Evaluated
  - Pros, cons and necessary forward work for each reference will be discussed
    1. Pre-filled/Sterile Liquid Nutrition
    2. Mixed powder/water Liquid Nutrition
    3. Solid Food Bar/Bites
    4. Helmet Feed Port

Bottom Line Up Front – November 2022 – See full report for details: "Recommendations for Developing Space Suit Integrated Food Systems and Delivering Nutrition Before, During, and After Lunar EVA"

BODY-17

Prefilled vs. Hydratable bags

Drink Bag
(Prefilled)

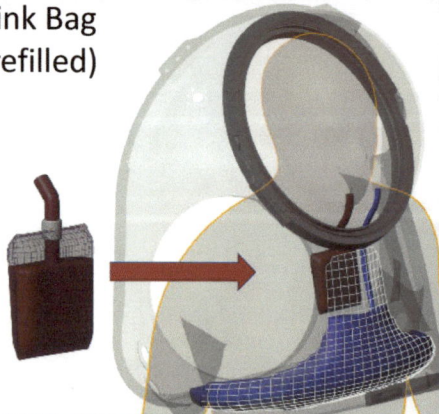

Drink Bag
(Powder)

Delivery Concept		In flight Prep	Special Capabilities Needed or Desirable	Advantages	Limitations
Drink Bag Prepackaged (Semi)Liquid	Thermostabilized	Open DURING EVA	Hands Free Access Hands Free opening/breaking seal	Ease of install/use/removal Consume when needed Variety of potential COTS	Higher launch mass (water) Delivery hardware has not been developed and feasibility is unknown
Drink Bag Powder	Rehydratable	Rehydrate BEFORE EVA	Hands Free Access Desirable: advances in engineering to extend food shelf life.	Low launch mass Ease of use/removal Variety potential COTS	Requires rehydration prior to donning pressure suit Consumed during early phase of EVA (within 2 hours of breaking seal/adding water)

Lichar Dillon | edgar.l.dillon@nasa.gov

Solid Bars

Delivery Concept		In flight Prep	Special Capabilities Needed or Desirable	Advantages	Limitations
Food Bar Solid	Low Moisture	Open BEFORE EVA	- Hands Free Access - Crumb prevention	- Low launch mass - Ease of install/use/removal	- Production of suitable food products (nonstick, crumb prevention, consistency, integrity, etc).

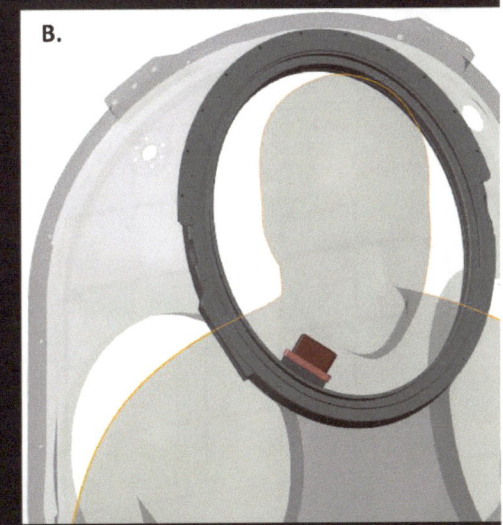

Helmet Feed Port

Delivery Concept	In flight Prep	Special Capabilities Needed or Desirable	Advantages	Limitations
Helmet Port (Semi)Liquid See Drink Bag	Install in Restrainer Pouch Stage before EVA	- Helmet Port - Vacuum tolerant packaging - Packaging that enables equilibration of pressure with suit to enable drinking - Potential need for refrigeration - Requires procedures and equipment for staging of food system external to the pressurized vehicle prior to EVA.	- No need for handsfree operation - Option to increase volume/frequency of nutrient intake without need for revisions of suit design. A.	- Complicated helmet port design/ concern for helmet port integrity - Need for specialized hardware - Requires use of gloved/pressurized hands - Requires staging of food for access to system external to vehicle. B.

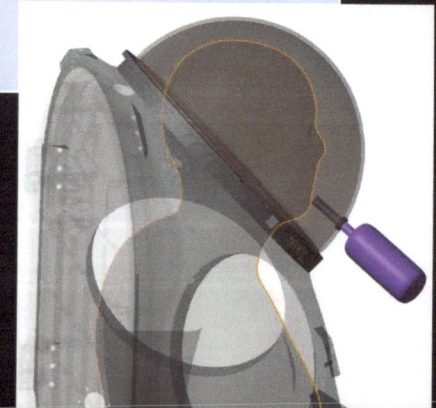

Suited Nutrition Standards

NASA-STD-3001 Volume 2

7.1.1.4   EVA Food Caloric Content

[V2 7004] For crewmembers performing EVA operations, the food system **shall** provide an additional 837 kJ (200 kcal) per EVA hour above nominal metabolic intake as defined by section 7.1.1.3, Food Caloric Content [V2 7003], in this NASA Technical Standard.

[Rationale: Additional energy and nutrients are necessary during EVA operations, as crewmember energy expenditure is greater during those activities. Consumption of an additional 837 kJ (200 kcal), similar in nutrient content to the rest of the diet, per hour of EVA would allow a crewmember to maintain lean body weight during the course of the mission. This is the metabolic energy replacement requirement for moderate to heavy EVA tasks.]

11.2.2.1   Suited Nutrition

[V2 11025] The system **shall** provide a means for crew nutrition in pressure suits designed for surface (e.g., moon or Mars) EVAs of more than 4 hours in duration or any suited activities greater than 12 hours in duration.

xEVAS Requirement

3.5.12   In-Suit Nutrition (RQMT-055)

The xEVA System spacesuit shall provide at least 1673.6 kJ (400 kcal) of crew nutrition while in a suited, pressurized configuration for EVAs of more than 4 hours duration, to be consumable at any point prior to suit doffing.

Rationale: Additional nutrients, including fluids, are necessary during suited operations as crewmember energy expenditure is greater during those activities. Nutritional supply during suited operations allows the crewmembers to maintain high performance levels throughout the duration of the EVA. Apollo astronauts strongly recommended the availability of a high-energy substance, either liquid or solid, for consumption during a surface EVA.

Consumption of additional nutrition that comes outside of xEVA System is expected to be consumed pre or post EVA. A total of 837 kJ (200kcal) per EVA hour must be provided to account for workload expenditures by a combination of EVA host vehicle and xEVA Suit System. This is the metabolic energy replacement requirement for moderate to heavy EVA tasks.

Applicability: ISS and Artemis